BASIC GUIDE BOOK ON WELDING ALUMINUM

Practical Techniques to welding mig and tig

Gael Ayden

Table of Contents

CHAPTER ONE 3

INTRODUCTION 3

CHAPTER TWO 25

PRE-WELD CHECKLIST FOR THE PROCESS 25

CHAPTER THREE 39

ADDITIONAL TIPS FOR ALUMINUM WELDING 39

CHAPTER ONE

INTRODUCTION

You probably know the look of an aluminum weld. An aluminum weld can look like a stack or dimes, and it is possible to create a consistent weld. You may have heard of the difficulties involved in welding aluminum. These are the questions we want to answer:

- What makes it so hard to make aluminum welds that are truly good?

- What makes welding aluminum so difficult?

This article will discuss aluminum welding. We will specifically focus on TIG welding and offer welding tips along the way.

Aluminum Welding Equipment

How to choose the right welder for your application

Since TIG welding aluminum is widely used for aluminum welding, this article focuses on it. It is important to realize that TIG welding doesn't have to be the only innovation in the process. Both TIG welding or MIG welding can achieve great welds within the aluminum material class.

TIG welding hardware and MIG welding hardware are very different. TIG welders use a tungsten terminal to keep the weld pool framed. The welding administrator also inserts a filler material into the puddle. The aluminum MIG welding hardware, which is specifically designed for this cycle, is interesting.

A spool gun is required to use a MIG welding machine to join aluminum materials. A spool weapon contains a loop of aluminum-filler material that is spooled into an cartridge. The handle of the MIG welding machine houses the cartridge. It is quite simple to explain that the aluminum filler wire needs to be accessed through the spool firearm.

The aluminum filler wire material is not as flexible as

ferrous filler materials. It can be pushed and pulled through twists of MIG links up to the handle. This article focuses on the TIG welding process of welding aluminum, as MIG spool firearms aren't used as often in this fledgling welding class.

When searching for a TIG welding professional, there are many options and highlights. Below you will see a selection of highlights that distinguish conservative TIG welders and better TIG welding.

Conservative TIG Welders

It is the following that makes a welding machine more cost-efficient:

- Cooling framework
- Maximum amperage range
- Duty cycle
- Accessories

For welding aluminum, TIG welders that are more efficient have certain prerequisites.

- Current AC capacity
- Cooling framework
- Ruggedness

These highlights will in any case be limited. It is perfectly acceptable to use a TIG welder with less highlights. It's all about how you use it.

How often will you weld aluminium

While you won't be able to weld every day or on multiple occasions per day for extended timeframes, it is possible to weld parts together from time-to-time. Conservative evaluation welders are always on the watch. Cooling is an essential part of any welder used for TIG welding.

Aluminum has a lower softening point than most ferrous metals so the welding amps must be higher. The warmth is more important when the amperage is higher. A cooling system with air is more practical for welders. They don't have any unusual methods of cooling TIG light.

A weld cannot be longer or more extensive. This would mean that the curve must be in existence for longer periods of time, generating a lot heat.

TIG welders that are efficient at welding thicker materials won't be able to reach the higher amperages required. Pre-warming the workpiece partially can help to reduce this problem. Welders with smaller welds are not equipped to handle the larger welds in thicker aluminum materials.

You should take breaks because of the heat involved in welding aluminum. Particularly when welding aluminum with a longer duration. You could endanger your gear by not doing so. For detailed obligation cycle assessments, consult your owner's manual. When you have to perform more welds, it is possible to arrange your obligation cycle too short.

The embellishments that accompany these welders are usually limited to:

- Standard welding light

- Standard foot pedal

- Regulator

Tip: Remember to include the cost of your argon-protecting gas container when calculating general expenses for welding gear. Accessibility in your area is key to estimating, but larger jugs may prove costly.

High Performance TIG Welders

The elite TIG welding machines for aluminum have everything that the more affordable evaluation welders do, but much more. Water cooling frameworks are used by elite TIG welders to reduce the production of heat in the light and the machine.

These cooling frames work in a similar way to your vehicle's cooling system. The cooling frameworks reduce heat development in the water and do not include a radiator. These machines have a wider amperage range and a longer obligation cycle. The welding machine has a heavier obligation part.

Often, embellishments that are included with higher quality machines have a better quality. You can take this example:

- Ergonomic TIG lamps
- Precision gas controllers
- Precision foot pedals with better goal

- Lighter extras and heavier obligation braces

This welding machine is not necessary for novice welders. This is especially true when you consider the large increase in cost. These welders won't make you worry about your finances, but they will increase your enjoyment of the welding experience.

Procedures and techniques for aluminum welding

Material preparation

It is essential to set up your material before welding. Clean your aluminum raw material for two reasons.

1. Aluminum forms an oxide layer on its exterior surface when it is chilled at the processing plants.

2. Cleansing oils and pollutants at a surface level by handling the material

Before you start your weld, there are two steps to clean the aluminum outside.

Step 1: Cleanse the exterior of the material from oils and other foreign substances. This is the most important step before you begin brushing. Brushing can cause more pollutants to enter the material. To clean the exterior of the material, you can use CH3)2CO. You can also use other surface cleaners specifically for aluminum readiness.

Stage 2 To remove aluminum oxides, brush the aluminum material's outside. This oxide layer can be easily removed using a brush. For this purpose, you should continue to use a steel brush. A carbon steel brush can only introduce ferrous toxins to your aluminum. The debasements can then cause weld quality problems. It is important to keep your aluminum clean. A well-maintained aluminum joint will make a big difference in the quality of your welding.

Tip Label your brushes with "Steel", "Pure," or "Aluminum" to avoid getting the brushes mixed up. It is best not to accidentally use a brush you normally use on carbon steel for your aluminum material. This can lead to a lower quality and degraded weld.

Tip Use homegrown (USA), dissolve source aluminum whenever your weld needs to be basic. Aluminum from countries other than the USA, especially China, can have contaminations that will result in poor weld quality and joint strengths. There will be a significant difference in the cost of aluminum from homegrown and imported countries. You will lose what you have saved in cash and modest materials. The majority of legitimate online and local material providers will have aluminum from home ready to go.

Machine Preparation

You should ensure that your welding machine has the right settings. Your machine's selector switch should be set to Alternating current (A/C). Your machine's upper range should have the amperage reach set to A/C. Your foot pedal will help you control this more precisely.

Because of its superior toughness and virtue, your TIG terminal must be made from pure tungsten. You should not sharpen the cathode as the rotating current can cause the tip of the anode to dampen and fall into your weld. For optimal weld execution and virtue, your protecting gas should not be altered by argon gas.

Tip: A weld setting that works well for someone else may not work for you. You can keep a "cheat sheet", which is a list of the best settings for welding, so that you can review your best welds. It is easy to tell when you have created a brilliant dot.

It will be a pleasure. Do not be afraid to go beyond the call of duty and save that setting for when you need it again. This is especially useful if you are sharing a machine with a friend. At that point, you will have the option of rehashing the wizardry.

CHAPTER TWO

PRE-WELD CHECKLIST FOR THE PROCESS

This is your "preflight" agenda. It's similar to how pilots make sure their plane is in good order before taking off. Are you sure that all of your equipment is ready for the welding activity? To ensure that everything is ready for the welding activity, you should create an agenda.

You should include the following things in your agenda:

- Check that there are no other objects around your workpiece that could catch fire.

- Determine your amperage

- Make sure your current is set at Alternating Current (A.C.).

- If your machine is water-cooled, make sure you check it for the most effective coolant.

- Secure all connections to your TIG light link, hose and hose

- Check the anode to ensure it is in a good condition and that it has as much "stick-out" as possible.

- Turn the valve on your jug to protect gas.

- Adjust your protecting gas controller for the maximum stream

- Switch on your auto-obscuring welder head protector and turn it off

- Always wear all of your PPE

Tip: There are many things you need to remember before starting your weld. It might be helpful to write your own list on paper and attach it to the highest point on your welder.

Ready Individual Protective Equipment.

Ensure that all of your Personal Protective Equipment is properly set up. Security is paramount in the field. Ensure that your PPE is always with you and your personal items before you begin welding. Your PPE basics for welding, as discussed in other articles:

- Safety glasses and Hearing protection
- Steel toe boots
- Protective cap for auto-obscuring welding
- Welding gloves,
- Leathers/fire-resistant coat
- Headwear that is flame-retardant

Although the above are essential for welding, you can choose to use the following PPE items to TIG weld aluminum.

- Safety glasses
- Cap for auto-obscuring welding
- Welding gloves.

While working in a welding shop, ensure that you always wear your safety glasses. Particularly when crushing TIG welding cathodes. After welding, tungsten material is often used to encapsulate TIG welding terminals. It is possible to have tungsten pieces fly through the air and crush your cathode.

Helmet Shading and Gloves

Adjust your auto-darkening helmet to the right settings to TIG weld aluminum. The amperage required to weld aluminum is usually higher than that used for other materials. To match the intensity of the light, increase the shade level. For instructions on how to adjust your welding helmet, please refer to your manual. It is crucial to choose and use the right type of welding gloves for TIG welding aluminium. To avoid any painful burns to your hands. The traditional TIG welding gloves are made of thinner calfskin gloves. However, you may want to use a thicker glove to weld aluminum.

Some welders prefer a thicker glove, similar to the "Mechanics", gloves. They are thicker and allow for more movement, mainly because they are made of thicker leather. You might have different preferences. You can always try another pair of gloves.

View the Weld Path and Angles

It is helpful to visualise the process of welding before you start. Particularly, the steps you need to take to ensure your weld is successful. You might find it helpful to close your eyes and imagine the angles at which your electrode, filler rod, electrode handle, and foot pedal should be placed throughout the welding process.

You might have a problem with your weld if you don't use this visualization method when you begin your weld. You might also find yourself using the wrong filler rod angle or electrode angle if you don’t use this visualization technique. Elite athletes use visualization techniques before the event starts. These techniques are used by elite athletes for the same reason as welders: to keep their focus on the event.

Create a weld, but don't dwell

To start your high-frequency arc, press your foot on your foot pedal. You should keep your eyes on the size and extent of the puddle on each side of your joint. You can control this using your torch angle. Start to create the proper sized puddle by dipping your filer into the puddle. Next, move your torch forward and continue to do so. Although you need to wait for the puddle under your welding torch to form, it is important to work quickly.

You will have trouble TIG welding aluminum if you don't feel the need to act quickly.

You can't make a puddle!

The Heat Zone is Out of Control

You can either bend in your workpiece, or blow through it completely.

Here is where the welding helmet's experience comes in handy. This is the ability to change your welding parameters on the fly.

Inspection (Pass/Fail).

A weld inspection is not something that beginners need to worry about. They should still be aware of the real-world consequences. Aluminum joints are used in extremely high-demand applications. Aluminum welds are used to hold together airplanes and race cars. These aluminum welds are carefully inspected for quality before approval.

The most advanced inspection methods include fluorescent penetrant and x-ray inspection. These inspections are used to check for cracks in weld. An easier approach for beginners is to check if there are any visible cracks in the weld, to verify consistency from start to finish and to determine if it is the right size. Don't worry if your welds fail to pass the mock inspection. Just start again until you are satisfied with your work.

Tip: If the weld cracks down the length of your bead it could be that your workpiece is getting too hot, or that your filler rod may not be the right match for your base material. This issue can be solved by either adjusting your amperage settings, or doing some research online to find the best filler for your aluminum alloy.

CHAPTER THREE

ADDITIONAL TIPS FOR ALUMINUM WELDING

Optimal conditions

Sometimes, your technique may be perfect but the environment around you is causing weld quality to drop. Wind is one environmental factor that could affect your weld quality. Your shielding gas may be leaking or not protecting your weld well if there is wind. The natural atmosphere that we breathe can cause porosity and unclean welding if there is no shielding gas around your weld pool. This applies to all welding technologies but is especially true for TIG welding, which involves welding non-ferrous materials like aluminum, titanium, or superalloys with high nickel content.

You can resolve this shielding gas problem by simply putting up screens to block wind and/or increasing the shielding gases output. Avoid blocking ventilation in the area you are welding in, as shielding gases and fumes can pose a danger.

Relax

Relax while you're welding. It is a complex process that requires skill and concentration. This can lead to stress and tension while you are creating your weld beads. You won't be able to weld more if you are tense after creating your first weld beam. Your subsequent weld beads may suffer.

You will see results if you practice. Relaxation allows you to focus more on your weld, filler dipping technique, as well as your heat affected area. If you get anxious about your weld, it will cause you to lose focus on your weld and filler dipping techniques will become less effective. Your heat affected zone will likely grow out of control.

Feel Comfortable

Being comfortable is key to focusing on making great welds. TIG welding allows the operator to sit down and position the workpiece. It is a great advantage to be able sit down and weld. A poor welding operator can often cause a weld to look bad. Poor welds can result from a distracted operator, such as a crooked wrist or a sore back.

Turn the workpiece and clamp it to a more comfortable place. Get a better cushion for your chair and ensure your foot pedal is in a good position.

Small adjustments can make a big difference in the quality of your welds. Sometimes, you may not be able to feel completely comfortable while you weld. This is especially true if you're welding from a fixed position. Try to be as comfortable as you can stacking as many variables in your favour. If you're welding in a bad position, it is a good idea to take a break. Weld quality can be affected by fatigued operators.

Request for Help

The welding industry is getting older and more people are retiring. There are many experienced welders out there who are willing to share their knowledge with anyone who is open to learning. Learning the welding process will eventually frustrate you.

You need to be able to use the many techniques and bits of experience that are available to you to master your trade.

Don't be discouraged if your aluminum welding bead is not perfect after three to four weeks of hard work. Reach out to experts and ask for assistance. You can find this help in many places: at a vocational school, community college or at work from your senior colleagues, as well as online videos and YouTube videos. There are many resources available. It is as simple as realizing you are not the only one struggling to learn TIG welding and asking for assistance.

Practice makes perfect

Although it may not sound obvious, this tip is very true in the case welding aluminum. Welding aluminum can be described as an art form. The machinery must have many parameters, and these parameters depend on various variables.

While you're welding, there are many conditions you need to be aware of. Experience and skill will help you fix any issues quickly. TIG welding aluminum is a complex process that requires precision. You must manage multiplc variables simultaneously in order to make a great weld.

Master welders will tell ya that once you think you know

it all, you are really not. While there is no substitute to experience in welding, education and practice can help you gain that valuable experience.

Conclusion

Welding aluminum is a difficult skill to master but it can be extremely rewarding. For skilled welders who are able to successfully weld aluminium, there are always job opportunities. These opportunities can be found in either the aerospace or shipbuilding industries. To improve your skills, you should learn to weld as often as possible. Also, be positive about the possibility of laying a great weld.

www.ingramcontent.com/pod-product-compliance
Ingram Content Group UK Ltd.
Pitfield, Milton Keynes, MK11 3LW, UK
UKHW021644190726
13853UKWH00001B/44